MIDDLESBROUGH'S
IRON & STEEL INDUSTRY

JOAN K. F. HEGGIE

'Born of iron, made of steel' – This sign and a sculpture by Stokesley artist Steve Tomlinson are at the entrance to the South Tees Business Centre in South Bank. (*J.H.*)

First published 2013

Amberley Publishing
The Hill, Stroud
Gloucestershire, GL5 4EP

www.amberley-books.com

Copyright © Joan K. F. Heggie, 2013

The right of Joan K. F. Heggie to be identified as the Author of this work has been asserted in accordance with the Copyrights, Designs and Patents Act 1988.

ISBN 978 1 4456 1283 6
E-Book 978 1 4456 1304 8

British Library Cataloguing in Publication Data.
A catalogue record for this book is available from the British Library.

Typeset in 10pt on 13pt Celeste.
Typesetting by Amberley Publishing.
Printed in the UK.

Introduction

Middlesbrough and Teesside are renowned for their association with the iron and steel industry. Iron has been produced there on an industrial scale since the 1850s, when Cleveland ironstone was discovered in the Eston Hills. Skilled workers and manual labourers alike migrated from all over the country to work in the many furnaces and mills which steadily overwhelmed the landscape. Any uninformed visitor could be forgiven for overlooking this heritage, however, as there are so few architectural reminders of that historical association; the furnaces and mills have been demolished, the corporate buildings are now nightclubs or pubs, the river is less polluted and the air is cleaner and quieter. Thankfully, some of the records of the iron and steel companies survive to remind us of that time in Middlesbrough's history.

The British Steel Collection (BSC) contains the business records of fifty iron and steel companies or affiliated organisations from the Teesside area. Held at Teesside Archives in Middlesbrough, the collection was donated by British Steel plc in the early 1990s but, due to a lack of resources, remained inaccessible to the public for over twenty years. From 2006 to 2011, Teesside University and Teesside Archives worked together to raise the funding needed to create an online catalogue and repackage the entire collection. As Project Manager, I worked with a skilled team to fulfil these aims and, in addition, to raise awareness about the contents of this significant local heritage resource. I also continue to work with a dedicated handful of volunteers, helping to make digital copies of the photographs and negatives contained in the collection. To date, over ten thousand images have been digitised. This book is an opportunity to share some of these wonderful pictures which, since the decline, demolition and regeneration of Middlesbrough's industrial landscape, provide a unique insight into the workplaces of the iron and steel industry, as well as the occasional glimpse of, what will be to many, familiar landmarks.

The growth and subsequent decline of the iron and steel industry is, in many ways, the history of Middlesbrough itself. Few towns have expanded so rapidly over such a short time as Middlesbrough, both in terms of population growth and economic output. In the late 1820s, Joseph Pease, a Quaker merchant from Darlington, instigated the extension of the fledgling Stockton & Darlington Railway (in which he had an interest) eastwards in search of deeper berths from which to ship coal from his mines in County Durham. Middlesbrough did not exist at that time, other than as a farm. Having established 'Port Darlington' (shown as Dent Wharf on later maps of Middlesbrough), Pease saw the investment potential of the area and persuaded others to join him in a company called The Owners of the Middlesbrough Estate. Together they purchased over five hundred acres of land, laid out a new town to the east of the railway tracks with a market square at its centre, and set about attracting new businesses to the area. In 1831 the population numbered 154; within a decade it had burgeoned to 5,463 and the town, sandwiched as it was between the river and the railway lines leading north and east to the newly built docks, was forced to expand southwards.

The discovery of ironstone in the Eston hills near Middlesbrough was the turning point for the growth of the iron and steel industry in the area. However, the importance of the river and the railways for transportation and the availability of coal from the Durham coalfields and limestone from Yorkshire cannot be ignored. In short, Middlesbrough was ideally placed to exploit the ironstone when it was discovered. Henry Bolckow and John Vaughan had set up an ironworks in Vulcan Street, Middlesbrough, in 1841 to work on finished iron, but this new discovery meant that they could produce it themselves locally. In 1852, they opened their first blast furnace, Eston Iron Works, in the area just south of Holme Beck Bridge and many others were quick to follow. In 1854, the Tees Side Iron Works was opened near Port Darlington and, across the river at Port Clarence, Bell Brothers established the Clarence Iron Works. To the east of the town, iron works were being built almost as quickly as the ink was drying on the agreements. Throughout the 1850s, blast furnaces were erected by a variety of different companies along the river from Middlesbrough to South Bank. The following decade saw the development of the land to the west of the Stockton & Darlington railway tracks in Middlesbrough, the area known as the Ironmasters' District, to accommodate this rapidly growing industry. By 1866, there were fifty-eight blast furnaces in Middlesbrough, Cargo Fleet, Eston (South Bank) and Port Clarence. In Middlesbrough itself, the Victorian ironmasters and their families were influential figures in all areas of civic life. Many were involved in local as well as national politics and encouraged the expansion of the town to include visible signs of Middlesbrough's success, such as the new Town Hall and Carnegie Library on Albert Road and the Royal Exchange in Exchange Square.

To separate Middlesbrough's companies and works from the rest proved to be impossible as their stories were so intertwined. The book is therefore organised into five sections which follow the River Tees from Newport in the south of the Ironmasters' District around the bend in the river to Port Clarence, North Ormesby, Cargo Fleet, South Bank, Grangetown and Eston. Each section begins with some contextual information about the iron and steel works in that geographical area, followed by a selection of images related to those works. The vast majority of the images are from the British Steel Collection and it has been a privilege to work with this element of the collection so closely for so long. Nevertheless, having such a rich resource to choose from has meant that the final selection was extremely difficult.

Most of the buildings associated with Middlesbrough's iron and steel industry, including the Royal Exchange, have been demolished, and new buildings, housing estates and roads now cover ground once criss-crossed with rail tracks, mills and foundries. It is incredibly difficult to visualise what the landscape was like when walking around the town today. These images trace the changing landscape from the latter half of the nineteenth century through time to the late 1960s when the industry was nationalised.

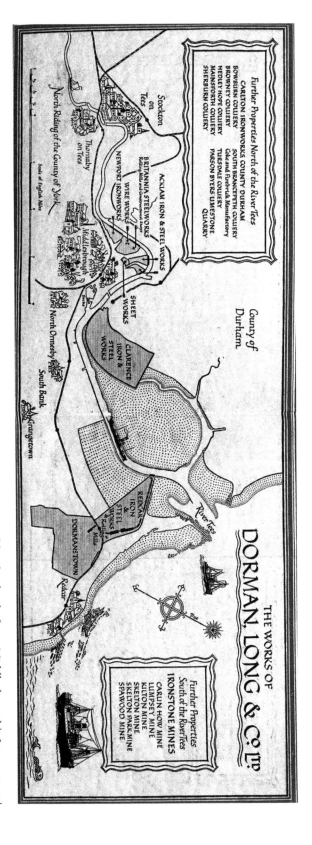

The Works of Dorman, Long & Co. Ltd. This map helps to illustrate the geographical area covered by this book, from Middlesbrough's Ironmasters' District to Grangetown and Eston. Originally produced by Dorman Long for a 1924 trade catalogue, it shows how diverse the company's interests were at this time. (*Bodleian Library, University of Oxford: John Johnson Collection: Ironmongery 9 (9)*)

Ironmaster's District

Although the Ironmasters' District is the area of Middlesbrough which most people consider to be the hub of its early industrial development, it is interesting to note that, on the 1853 map, it did not exist. The land west of the 'Old Branch' railway tracks was described as the Newport Marshes and contained only a solitary shepherd's house and several sheep pens. The early representatives of what was soon to become the dominant industry were the Tees Side Iron Works near Port Darlington (Snowdon & Hopkins), Bolckow & Vaughan's Middlesbrough Iron Works on Vulcan Street, the Tees Iron Works at the entrance to Middlesbrough Dock (Gilkes, Wilson, Pease, Leatham & Co.) and the Ormesby Works, owned by Cochrane & Co. Further east was Bolckow & Vaughan's newly opened Eston Iron Works, situated at the northern end of the Eston Branch Railway which conveyed the ironstone from their mines. Only farmland and pastures separated Cochrane & Co. and the Eston Iron Works. A decade later, however, at the time of Chancellor of the Exchequer William Gladstone's 1862 visit to Middlesbrough, '... this remarkable place – the youngest child ... of England's enterprise'[1], the Ironmasters' District was under development.

By 1882, it was well-established as an iron smelting and manufacturing area, although diverse in respect of the number of owners. Statistics show, in addition to pig iron production, 852,000 tons of manufactured iron was made in Cleveland that year, the highest for any region. Steel production, on the other hand, was growing at a much slower rate than the other regions (e.g. Scotland and South Wales). From Newport, moving north through the district, the layout was as follows: Newport Iron Works (B. Samuelson & Co.); Newport Rolling Mills (Fox, Head & Co.); West Marsh Iron Works (Dorman, Long & Co.); Britannia Works (originally laid down by B. Samuelson & Co. but by this time owned by Dorman, Long & Co.); Ayresome Iron Works (Gjers, Mills & Co.); North Eastern Steel Works (North Eastern Steel Co.); Acklam Iron Works (Stevenson, Jacques & Co.); Linthorpe Iron Works (E. Williams); Roseberry Steel Works (Butler Bros.); Ayrton Rolling Mill (Jones Bros. & Co.); Acklam Iron Foundry (James Ritchie); and various other foundries and related businesses. Tees Side Iron Works (by now Hopkins, Gilkes & Co.) and the Middlesbrough Iron Works (Bolckow, Vaughan & Co. Ltd) were still operating and Raylton Dixon had opened his shipyard (later Cleveland Dockyard).

The next fifty years were to change the layout of the Ironmasters' District as certain companies grew in stature and reputation and began to dominate the industry. Since taking over the West Marsh Iron Co., in 1876, Dorman & Long had gradually increased their activities, firstly by leasing additional space from neighbouring works (e.g. Britannia, which they subsequently purchased in 1882), and then by taking over other companies. In 1889 the company was incorporated (Dorman, Long & Co. Ltd) and, a decade later, took over Ayrton Sheet Mills and a wire works to the south of West Marsh. In the same year, they started the process of taking over Bell Brothers Ltd, of Port Clarence. In 1903, the company purchased the North Eastern Steel Co., which by then owned its neighbour, the Acklam Iron Works. In 1917, DL & Co. Ltd purchased Sir B. Samuelson & Co., thereby gaining control of the Newport Iron Works. In 1923, the various names under which the companies had continued to trade were abandoned in favour of Dorman, Long & Co. Ltd. With the exception of Ayresome Works, which continued to be run by Gjers, Mills & Co., the vast majority of the district was swallowed up into the Dorman Long organisation. From the 1930s until nationalisation in 1967, the company gradually relocated much of its operational activities to their Clay Lane and Redcar sites, leaving an industrial wasteland behind. Since the late 1970s, the area has been known as the Riverside Park Industrial Estate and, after years of investment and regeneration, is home to industry once more.

1. "Mr Gladstone at Middlesbro'." *Leeds Mercury* (Leeds, England), 10 October 1862. 19th Century British Library Newspapers.

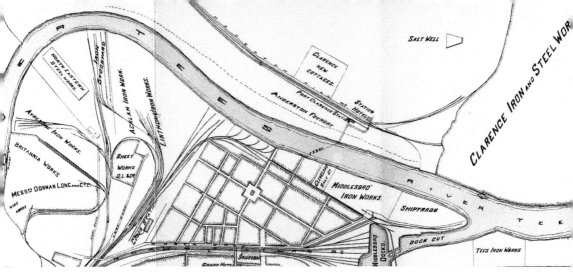

Ironmasters' District, Middlesbrough. This map was published as part of a company history in 1901, shortly after Dorman Long merged with Bell Brothers, based on the north bank of the Tees at Port Clarence. In addition to the West Marsh Works, the company had a wire works, rolling mills, a steel works at the neighbouring Britannia Works (formerly Samuelson's) and a sheet mill, which they had purchased in 1896. The Cart Road near the Goods Station is now known as Forty Foot Road (see p. 43). (*Teesside Archives BS.DL.7.5.2.1*)

This aerial photograph of the Ironmasters' District is thought to date from the 1940s or early 1950s. The junction of Forty Foot Road with North Road can be seen at the bottom. The Britannia Works of Dorman, Long & Co. Ltd has become a centre for constructional work, with the iron and steel being supplied from the company's furnaces. The Test House (still standing, see p. 43) was used to test the steel for the Sydney Harbour Bridge and can be seen just above the fork in the roads. The white smoke shields the view of the Ayresome Iron Works (Gjers, Mill & Co.); both the North Eastern Steel Works and Acklam Iron Works have become part of the Dorman Long group. On the north bank of the river is the Furness Shipbuilding Company Ltd of Haverton Hill. (© *English Heritage (Aerofilms Collection); 2571*)

Newport

Above: Records from as early as the twelfth century mention a ferry landing at Newport. This drawing, entitled 'The Bend of the River – Newport', is thought to date from about 1880. The Newport Iron Works was established in 1864 by Bernhard Samuelson and this image depicts an early view looking north towards Middlesbrough and Haverton Hill. The wharf was small but the railway wagons and crane show that it was used for freight. The footbridge, which allowed people to cross the tracks of the Stockton and Darlington Railway, was situated at the western end of Samuelson Street. (*1506*)

Left: This sign was at Dorman Long's Newport Iron Works - Central Research Laboratory which was located immediately south of the Newport Bridge approach road (see page 11). (*5267*)

Newport Iron Works, Middlesbrough

In 1917, Dorman Long purchased Sir B. Samuelson & Co. and gained control of the Newport Iron Works, although the name did not change until 1923. In front of the furnaces, 'pig beds' were laid out in sand ready to receive the molten iron when the blast furnace was 'tapped'. The author of the postcard, a man called Charlie, was not impressed. He wrote: 'Just a couple of views of the place, it is not up to much'. (*Valentine's "Phototone" Series - J.H.*)

A boiler explosion at the Newport Iron Works in April 1927 caused considerable damage but, thankfully, no loss of life. According to *The Times* newspaper, the force of the explosion caused the boiler to land over one hundred yards away, an iron chimney stack to collapse and windows of nearby properties to shatter. What is left of the boiler can be seen near the rail tracks to the top right of the picture. (*2269*)

Building a second bridge across the Tees was thought necessary as early as 1920, just nine years after the opening of the Transporter Bridge. However, it took four years to decide upon the location (Newport) and a further six years before the plans were given Royal Assent. Dorman, Long & Co. Ltd.'s tender for a vertical lift bridge was accepted and work began in 1931. Each section was fabricated and test-assembled in Dorman Long's Bridge Yard at Britannia before being transported to the site. This is the lifting span for the Durham side of the bridge, taken in 1932. (2542)

At the time of its construction, the Tees (Newport) Bridge was the largest vertical lift bridge in the world and the first to be built in Great Britain. The bridge was 250 feet wide and contained 6,500 tons of steel. To facilitate river traffic, the road deck could rise to 120 feet to allow ships to pass below. Although much needed, the bridge was not without controversy. Over sixty houses in Newport were demolished to make way for the approach road, resulting in two hundred and ninety four people being re-housed in the Whinney Banks estate. (2217)

The Newport Bridge was formally opened by the Duke and Duchess of York (later King George VI and Queen Elizabeth) on the 28th February 1934. Streets nearby were decorated with bunting; local people and hundreds of school children crowded the pavements to wave at the royal couple as they passed. This image shows the bridge deck in the raised position and the industrial landscape of Haverton Hill in the distance. Bottom right is Dorman Long's Central Research Laboratory. (*Teesside Archives US 96-2*)

Above: The Central Research Laboratory and offices from the south east, 1950, showing the east bank tower structure of the Newport Bridge. Pedestrian access to the laboratory was gained via an iron staircase from the bridge approach road. It cannot have been a very peaceful place to work with the railway tracks immediately behind the building, shipping traffic on the river to the front and the road bridge almost within touching distance! (*5694*)

Right: Looking up river towards the Newport Bridge, an incredible feat of engineering for its time. The road deck was raised and lowered using two electric motors but a back-up engine (petrol) was also installed in the winding house structure, seen here at the midpoint of the bridge. In the event of that failing, it was actually possible to raise the span by hand as it was so finely balanced. In 1990, the road deck was pinned down permanently due to the decrease in river traffic. (*1404*)

Wire Works

The drawing blocks in the Wire Works, dating from before 1901. The Wire Works were situated next to the West Marsh and Britannia Works and were inter-connected by rail tracks. In the 1920s, it took just over a minute to roll a steel billet measuring 28ft x 2in square into a steel rod measuring three quarters of a mile at 3/16 diameter. (*8962*)

The drawing blocks were used to achieve the correct diameter or thickness of wire. As the wire was passed through a gauge, it was squeezed to the correct thickness, thereby increasing its length. The wire was held on one spool, drawn through the machine and coiled on a corresponding spool. (*1313*)

Above: To prevent corrosion, the wire was galvanised by passing it through a bath of molten zinc. Probably dating from the 1930s, this machine is feeding wires from several spools through a mechanism to keep each wire separate from each other. This allowed the wire to be coated evenly with the liquid zinc. (9002)

Right: This workman is using a tool or brush to ensure the zinc coating is applied evenly. Just a year after this was taken, the company installed a new 'hot-dip' galvanising system with a re-winding frame. (5324)

13

The Old Rod Mill in 1958. The workman (or handler) passed the rod through the roughing mill drives, where it was collected, looped back on itself and fed back through by another handler. The looping floor, often made of cast iron, can be seen in the foreground. This process continued until the rod was the correct thickness. (*5341*)

A closer view of the roughing mill drives in the Old Rod Mill, 1958. A roughing mill is the first stage of rolling which converts steel ingots into other forms. The toothed drives were often cast from bronze. An incredible amount of force is required to compress the rods down to the correct size which means the drives have to generate high torque. (*5314*)

Right: These spools or bobbins holding the wire were called 'double-headed swifts' and were used to automatically feed wire into the galvanising line. The reel of wire on the bottom of the swift was drawn into the galvanising process. The end of this reel was attached to the coil above it. As the lower reel reached its end, the upper coil dropped to the bottom enabling another reel to be loaded onto the swift. (*5325*)

Below: Dating from the early 1960s, the next three pictures illustrate the process of transporting coils of wire to Middlesbrough Dock for shipment. These 8ft coils of 0.276in diameter high tensile steel wire were loaded onto a lorry using a hoist as the coils weighed 10 cwt (approximately 500 kilos). (*10165*)

Above: The lorry driver, from D. Tarren Ltd., of Thornaby on Tees, checked the load as he pulled out of the loading bay. The coils were not tied down; presumably the weight of the wire was thought to be sufficient to prevent slippage. The yards attached to the Britannia Works can be seen in the background. Also in view are Locomotive No. 34 and a jib crane mounted on a bogie bearing the logo of DL & Co. Ltd. (*10167*)

Left: At Middlesbrough Dock, the coils of wire were hoisted on board ship bound for Port Swetterham, Malaya. (*10169*)

Britannia

Britannia Works. A familiar sight to those who worked at the Britannia site, either in the Steel Works, the Mills or the Bridge and Constructional side of the business. A series of photographs entitled 'Where We Work' was taken by the company in the 1950s. Several of them are featured in this book. (5258)

Britannia Works, September 1929. This view from the north bank of the River Tees shows the long sheds of the Britannia Works, the blast furnaces of Ayresome Iron Works and the Furness shipyards on the north bank at the bend of the river. Just in view on the far right are the furnaces of the Acklam Iron Works. (1589)

Draughtsmen in Britannia's Drawing Office above the Test House, September 1945. Plenty of natural light was needed but difficult to achieve given the air pollution caused by industry. Overhead lights were also suspended overhead. Each man had a desk of his own, one side with an adjustable easel which was used for working on individual drawings; the other holding the stack of drawings being worked on. Surprisingly, considering the date of this image, the men appear to be from a wide age range. In contrast, the two apprentices in the Blacksmith's shop at Britannia Works seem very young. (*9493 & 9504*)

Workers in the Pattern Shop, Britannia Works in 1951. Using drawings as a guide, the men created wooden replicas of objects or 'patterns' for use in the foundry. Patterns are used in casting metal; fine sand is forced around the pattern in a 2-piece mould. When the mould is separated, channels are made in the sand to let the hot metal in and the air out. The wooden pattern is then removed before the mould is put back together again, leaving a cavity into which the molten metal is poured. (9516)

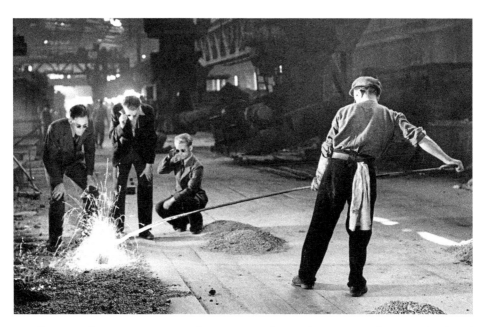

Metallurgy Students at Britannia Works in 1949 watching a furnaceman take a sample of steel from an open hearth furnace. The students, who were spending time in industry as part of a Sandwich Course, were wearing protective glasses to view the hot metal through, as was the furnaceman. The machines in the background were chargers for the steel furnaces. (5290)

Left: The last official cast of steel at the Britannia Steel Works in November 1953. New, high capacity blast and steel furnaces were built at Clay Lane (Cleveland Works) in the 1950s and the older furnaces in Middlesbrough were gradually closed. Hardly any of the men were wearing protective eye glasses, although they were in use at the time. In the centre of the picture is one of the two large hooks which lifted the steel ladle; in the foreground are empty ingot moulds. (*9545*)

Below: The last official cast of steel at the Britannia Works in November 1953, showing the steel ladle in a suspended position. In this image, the foreman is using the handle to release the metal from the bottom of the ladle into the waiting ingot moulds. (*9546*)

In the early 1960s, an Apprentice Training School was built in the shadow of the Test House at Britannia. This was one of several facilities in the area (another was opened at Dorman Long's Clay Lane site in 1962). Trainees received classroom lectures and skills training to enable them to use equipment and machinery similar to that found in the workplace. (*9609*)

The class of '61. Twenty-two young apprentices attended the Drawing School at the Britannia Works in 1961. Are you in this group, or do you know the names of these young men? No-one thought to write the names on the print at the time, unfortunately. The posters hanging on the walls are familiar. Similar images appeared in *Dorman Long Illustrated* (the company magazine). (*9610*)

Bridge & Constructional

By the early 1920s, Dorman Long were defining themselves as specialists in the fabrication of structural steel for bridges and buildings. They invested heavily in the constructional workshops in Middlesbrough, filling them with the latest tools and machinery, such as radial drills (shown here). They were laid out in a similar way to a modern day assembly line, so that as each operation was completed in turn, the item would move forward to the next stage. Rail tracks were laid between the stations to move sections through the workshop with the minimum of handling. Note how much cleaner these works look compared to previous images. (2511)

Just one example of the types of building erected in the post-World War One period for which the Bridge and Constructional Works supplied steel beams. 2,300 tons of structural steel was supplied for Bush House in The Strand, London. The building was erected for the Bush Terminus Company of London and New York and was part of the regeneration of the Strand area. The building was faced with Portland stone and the shape of the portico can be clearly seen in this image. In the early 1940s, the BBC moved into Bush House and remained there until 2012. (1952)

A view of one of the Wellman 5 Ton 'Creeper' cranes used by Dorman Long in the building of the Sydney Harbour Bridge in the 1930s. This undated picture was taken in the bridge yard. It is possible that the crane was brought back after the bridge was completed and used but it is more likely that the crane was assembled and tested before being shipped to Australia. (*1286*)

One of three welded bunkers manufactured in 1936 by Dorman Long for Imperial Chemical Industries (ICI) at Billingham. The bunker was transported to the site by T. O. Harrison Transport Ltd., of Middlesbrough. Harrison's used to have a large advert on the side of a building near Ormesby Crossing. (*2304*)

This 98 ton girder was fabricated for the Cumberland Hotel, Marble Arch, London, which opened in the 1920s. One reason why the man was standing next to the girder was to help the viewer establish the scale of the structure. This old chap, or someone of a very similar age, appears in quite a few photographs! (*8951*)

Two large girders for the Olympia Footbridge at Kensington station leaving Bridge & Constructional's site in 1929 on low bogies pulled by a steam locomotive. This bridge was eighty four feet long in total and long enough to cross all four rail lines at Olympia. The lattice structure of a travelling crane can be seen just above the roof tops. (*8771*)

An image from about 1930 showing a young boy passing hot rivets to a workman who then used the hydraulic gun to fix it in place in the girder. Behind the boy a different type of riveting machine can be seen. This was used when both sides of the metal could be inserted into the jaws of the riveter (see page 66). (*8806*)

A shore span for the Lambeth Bridge, London, which was built and erected by Dorman Long. This elevated view was taken in the Bridge Yard in Middlesbrough in 1930. The steel for each section was produced in Dorman Long's furnaces and then fabricated in Bridge & Constructional's workshops. Pictures like this were taken to provide evidence of progress to the client, in this case London County Council, often in advance of requesting the next staged payment. (*2532*)

A navigation span highway girder for the Storstrom Bridge, Denmark, fabricated and constructed by Dorman Long. A combined road and rail bridge, it was opened in 1937. It is almost 2 miles long, has 49 piers, 50 spans and 3 central arches. Recently it was announced that it would be replaced within the next decade. This image demonstrates not only how huge this individual span was, but also how large the bridge yard had to be to accommodate it. Look closely, there's a man standing on the girder to illustrate the scale. (*1125*)

The caption on the reverse of this image reads: 'When Middlesbrough people visit London they may well be proud of these evidences of the townsmen's efficiency as expressed in these three magnificent examples of steelwork as applied to the most modern forms of architecture'. The image must date from after 1932 when Lambeth Bridge was opened. Dorman Long produced 11,500 tons of steelwork for Thames House (left of picture), 6,500 tons for Imperial Chemical House (right) and 4,620 tons for the Lambeth Bridge. (*1931*)

Sydney Harbour Bridge, Australia, opened in 1932. This arch bridge, with granite piers and granite pylons, was the design selected out of the seven tenders submitted by Dorman Long to the New South Wales Government. The steel was produced in Middlesbrough and then shipped to Australia where it was fabricated in workshops constructed on site. The arch was created by cantilevering the structure out from both sides of the harbour and meeting in the middle. It took eight years to complete and its origin remains a great source of pride for Teessiders. (*1605*)

Ayresome

The blast furnaces of Ayresome Iron Works, owned by Gjers Mills & Co., and built in 1871. These furnaces had to be manually charged by loading barrows which were then hoisted to the charging platform at the top of the furnace. The hoist and charging platform can be seen in image 1594. These pictures are undated but are thought to have been taken in the late 1960s when the furnaces were demolished. (*1594 (top) & 1595*)

In 1896, the Acklam furnaces became part of the North Eastern Steel Company which, in turn, became part of Dorman Long in the early 1900s. By the time Dorman Long published their trade catalogue in 1924, the furnaces had been part of the firm for twenty years. (5265)

Three furnaces were originally built by Stevenson, Jacques and Co. in 1865, with a fourth being added within a few years. This postcard dates from between that date and 1912, when it was posted to a Mrs J. Walker in Coalbrookdale. A hoist can be seen behind the furnaces to take materials up to the charging platform. (*Valentine's Series – J.H.*)

Acklam Ironworks showing the pig beds in front of the four blast furnaces in varying stages of readiness for the molten iron. When the furnace was tapped, liquid iron would run into moulds set out in sand. The moulds were supplied from a main furrow and, because of the resemblance to piglets suckling at their mother, the ingots became commonly known as 'pigs' and iron as 'pig iron'. (2573)

Thought to be Acklam Works in the 1930s. The descriptions reads: '...ore, coke and lime-stone [are] mechanically fed at regular intervals at the top, these three constituents being known as the 'burden'. In the reduction of the ore, coke is used as the fuel and limestone as the 'flux'. The illustration shows molten basic iron flowing from the tap hole at the foot of the furnace into a bogie-ladle for conveyance to the Smelting Shop for conversion into Steel'. (8829)

Acklam Works, Connal's Pig Beds. Taken in 1954, this picture shows a more modern method of laying out pig beds. A Ransome tractor with 'shapers' is used to make regular impressions in the sand. Presumably the rake attachment, held by the workman, was for raking the beds once the pigs had been removed. (10247)

Above: Acklam Works, 1954 - The huge ladles of iron were transported from the furnace by rail (top) to the pig beds below. The cross-frame supported the ladle as it was tipped forwards, allowing the white hot contents to pour into the pig beds. Also seen is waste by-product (slag) which solidified at a lower temperature than the iron. (*10248*)

Right: Acklam Steel Works. The iron destined to become steel passed through mixer furnaces in the Smelting Shop before being poured into Steel Furnaces for the process of oxidation. Each batch was poured into the furnace which had been 'charged' or heated up to the required temperature. The process of removing the carbon from the iron took several hours but it was easier to achieve the correct carbon content using open hearth furnaces (as shown here). The furnace was then tapped and the steel poured into ingots (see p.20). (*8805*)

These men are charging an open hearth furnace at Acklam Steel Works ('D' furnace). Part of the process involved adding scrap metal, or the slag-forming materials such as limestone. Protective glasses were more commonly worn by this time (1950s), as was a long apron to prevent sparks burning their clothes; it was hot and dirty work. (5291)

Steel for the rolling mills had to be of a uniform temperature to facilitate the necessary 'plasticity' required. Ingots were placed in soaking pits to achieve this. Contrary to its name, there was no water involved! Soaking pits were insulated, temperature-controlled chambers, normally located underground. The steel ingots were deposited or removed (as shown here) from above via an access hatch. (10223)

The ingot was then 'hot rolled' in the roughing mill to start the process of shaping it to the desired size. This image, dating from 1948, shows two ingots passing through the roller at different stages in the process. The ingot on the right is moving through the wider roll, while the one on the left is being channeled through a narrower roll, which forces the metal to elongate. The ingots were passed repeatedly through the rollers until the correct size was achieved. (*10225*)

Different products, such as bars, were rolled using smaller machines and had to be passed back and forth through the rollers. Here the workman has gripped the bar with pincers and the second man was helping to either lift the bar into the rolling machine, or remove the bar from the machine. (*10236*)

33

Slab steel was rolled into different thicknesses of plate or sheet steel, depending on its intended end use and the customer's requirements. The slab was squeezed between rollers which rotated in opposite directions. The quality of the steel was regularly tested. This image shows lengths of steel being assessed for straightness (note the pegs and string method), prior to 'bending tests' being carried out in 1935 at Acklam Steel Works. (2598)

Acklam Coke Ovens, 1923. Most iron works had coke ovens on site to produce this vital material for the furnaces. This is a rare image of Acklam's coke ovens which were located to the east of the blast furnaces. In the background are rail wagons bearing the N.E. mark of the North Eastern Railway company. (2569)

These two photographs from the 1950s show slag pots being transported from the furnaces to the slag and ballast disposal area. A locomotive pulls several of these specially designed pots at once. The jib crane, also on rails, tips each pot in turn and the waste material is deposited on the ground. This material was often still extremely hot, or even in semi-liquid form (see p.87). Slag was a by-product of the process that was often used as an aggregate in land reclamation or the construction business. Many of the pots used in the Teesside works were made by Ashmore, Benson & Pease of Bowesfield Lane in Stockton. A surviving example can be seen in the grounds of the South Tees Business Centre at South Bank. (*10260*)

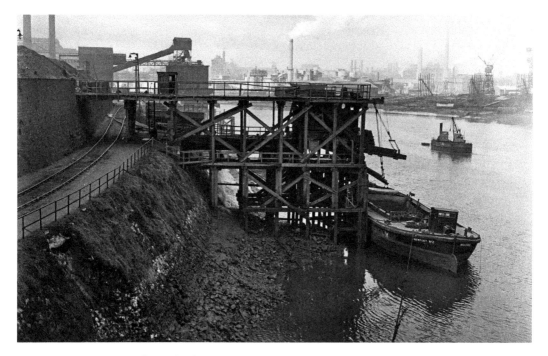

Acklam Works – No. 1 Slag Wharf, 1953 at low tide. The vessel in the water below the wharf was called *Newport No. 2*. There were several of these boats at Newport and Acklam Works (see below). On the other side of the river, industry is growing. In view is the cement works. (*10246*)

Acklam Works – Riverside Pump House, 1956. The boat tied up alongside is *No. 1 Acklam*. In the background hoppers are being moved by an elevated pulley system, similar to a ski lift. Can anyone remember where these were and what they were used for? (*10263*)

The Ayrton Sheet Works was acquired by Dorman Long in 1899 and this photograph was taken not long afterwards. The large fly-wheel engine in the middle of the floor powered the presses which the men used to flatten and trim the 'black sheets' (un-galvanized) before stacking. (*5266 & 5837*)

In this interior shot of the works, taken in the 1930s, three different types of sheet can be seen – flat, troughed and corrugated. The men used stencils to mark each sheet, presumably with the manufacturer's name or the customer's details. The worker's uniform of this period was trousers, shirt and waistcoat, an apron and a flat cap. Sometimes a piece of material was worn around the neck to absorb sweat. (*8792*)

The Sheet Mills were remodeled in the 1920s and new equipment installed. This picture, taken in 1939 at Ayrton, is described as a 'mechanical mill' which appears to be a form of finishing mill, or a means to cut the steel into smaller strips. (*5792*)

Right: No. 3 High Mill, Ayrton Works, 1947. A high mill was one which contained two or more rolls through which the hot metal was passed. The greater number of rolls enabled the steel to be passed in one direction and returned through the additional rolls. The operator was controlling the direction of the rolls, while the dials indicated that equal pressure was being applied across the plate. (*10026*)

Below: Corrugated sheets were galvanized sheet steel which had been through the hot-dip process described previously (see p. 13). To shape the corrugated sheets, workers passed each one back and forth through special presses. There is a clearer view here of the markings on some of the sheets. The trough steel (lower right) shows the Order No (O/2210A) and its destination – Copenhagen. (*2568*)

Corrugated sheets being loaded into rail wagons in the late 1940s for transportation to the docks and onward shipment to the customer. (*5810*)

During World War 2, Ayrton Works was involved in making air raid shelters from corrugated sheeting and concrete. The corrugated sheeting formed the interior surface (seen here) and gave the structure the domed shape. The structure was partially embedded in the ground and the concrete outer surface added. The blast furnaces of the Acklam Iron Works can be seen in the background. (*5696*)

Female workers were employed by many of the iron and steel companies during the world wars, primarily to replace men who had been called up into active service. Many were retained afterwards and worked in many roles, such as welders, crane drivers, labourers and machine operatives. In images dating from 1955, these women were operating Rhodes Plate Presses (manufactured by Joseph Rhodes Limited of Wakefield). In the first picture, a central hole was punched into circular steel discs or 'buffer plates'. In the second, the two women on the right were creating circular blanks out of sheets of steel, while the third woman (left) punched the central hole. The finished buffer plates can be seen on the table. (*5828 & 9982*)

Other products made at Ayrton Sheet Works and their uses. The 'N' line double decker bus to Berwick Hills next to a bus stop manufactured with sheet steel from Ayrton. The location is believed to be Overdale Road, just before the turn into Crossfell Road. Does anyone recognize any of the passengers? The second image is of plastic-coated steel sheeting being fixed to the roof of the supporters stand at Middlesbrough Football Club's Ayresome Park in 1962. Plastic coating enabled sheeting to be produced in different colours. Ayresome Park's sheeting was bright red to match the club colours. (*5799 & 5934*)

FORTY FOOT RD. TS2

A modern day view of the Ironmasters' District, which now houses the Riverside Park Industrial Estate. Many of the businesses located here are still involved with metal working, supplying steel and engineering. The first image shows the rail tracks running north and Forty Foot Road (for comparison, see p.2). Taken from the bridge over the rail tracks, the William Lane Brass Foundry is the first building on the left, with Parson& Crosland behind. The view of the Test House (to the left) is masked by trees, but can be seen in the second image. (*J.H.*)

Town and Docks

The Middlesbrough Exchange Company Limited was formed in 1864 to encourage the development of the iron and steel trade. The impressive Royal Exchange building, which filled the space between Wilson Street and Zetland Road, was opened in 1868. An iron market took place in the huge vaulted hall every Tuesday and Friday. There was also a gentlemen's club and other offices in the building which were rented out to local businesses. The new railway station was built across the road and the major banks all had equally impressive buildings nearby. The Post Office (now Teesside Archives) was ideally placed for company business. The seal is now outside Steel House in Redcar. (1627)

In 1946, Dorman Long moved into the Royal Exchange Building and used the vaulted hall as the Drawing Office (see p. 45). As can be seen, Middlesbrough Bus Station was located just behind the Dorman Long entrance, at the junction of Wilson Street and Marton Road. (5455)

Two interior views of Dorman Long's Drawing Office in the Royal Exchange building which illustrate some of the grandeur of what would have been the iron and steel market hall. In the 1950s, however, when these pictures were taken, additional strip lighting has been suspended above the draughtsmen's tables and about fifty to seventy desks crammed into the space. (*5458 & 5459*)

In the 1980s, the Royal Exchange was demolished to make way for the A66. The bus station was relocated and the top end of Marton Road cut off from the rest. This view, looking west from approximately where the old bus stands would have been, shows the A66 with a nightclub underneath. The statue is of Henry Bolckow which was relocated to Exchange Square in 1986. The carved relief stones (see inset), which once graced each ground floor window, are the only clues of its former existence. (J.H.)

An aerial view of the shipyards of Sir Raylton Dixon & Co. Ltd., and the entrance to the Middlesbrough Dock taken in 1921. The dock, and the area around it, was a hive of activity. The little footbridge in the centre of the image is still there (see p. 47) and St John's Church on Marton Road can be seen in the distance, as can the rows of terraced housing in that area. The area to the left of the footbridge, once covered with the industrial buildings of Tees Iron Works, is now home to the Riverside Stadium. (10360)

Looking west from Cochrane's Wharf to the iconic Tees Transporter Bridge, 1930s. The river was busy; the coaling plant of Port Clarence can be seen on the right and the entrance to the Middlesbrough Dock (Dock Point) is just visible on the left. Opened in 1911, the Transporter bridge was designed by Cleveland Bridge & Engineering Ltd., of Darlington and erected by Sir William Arrol & Co., of Glasgow. At that time, Dorman Long had not yet entered the bridge-building side of the business. (626)

A modern view of 'Middlehaven', as Middlesbrough Docks are now known, after years of regeneration. This picture is taken from the footbridge (see p. 46) and looks north-west past the Temenos sculpture and the Dock Clock towards the Transporter Bridge. Middlesbrough College is out of shot, to the left of the Dock Clock; Riverside Stadium is behind the photographer. (J.H.)

Port Clarence

The Bell family was originally associated with the Walker Iron Works of Tyneside, the firm which had employed John Vaughan (later the co-founder of Bolckow & Vaughan). The eldest son, Isaac Lowthian Bell, was educated in metallurgy, physics and chemistry and used his skills in the family firm. In 1844, he formed Bell Brothers with his younger brothers, Thomas and John. At first they leased a furnace at Wylam but, hearing of Vaughan's success in using Cleveland ironstone, the company entered into a lease with the Ward Jackson family of the Normanby Estate to mine for ironstone under their land. As a condition of that lease, Bell Brothers were required to build their ironworks on the north bank of the river. The works were served by the Clarence Railway, a branch line of the West Hartlepool Harbour & Railway Company (Chairman: Ralph Ward Jackson), the rival company of the Stockton & Darlington Railway. Bell Brothers 'blew in' their first three blast furnaces in 1854 and had doubled their iron production in less than two years and, by 1898, was producing over 300,000 tons of pig iron per annum.

Bell Brothers was a scientifically innovative company and, under the guidance of Lowthian Bell, it developed a process for making steel from Cleveland iron. The company developed a method of exploiting the salt seam under their land in the 1880s. It also had a track record of successfully exploiting the many by-products created by the iron and steel making process, such as coke, coal gas, tar and benzol, and had opened a refinery plant. With an extensive portfolio of collieries, a limestone quarry and several ironstone mines, it was an attractive company. In 1899, following a slump in the demand for iron and several years without making a profit, it merged with Dorman, Long & Co. Ltd, although it continued to trade under its own name until 1923. A year later, Dorman Long reported that there were eleven blast furnaces, nine open hearth steel furnaces, a 300-ton mixer furnace and two mills at the Clarence Works. A new coaling plant was opened in the 1920s (see p. 54) and the site continued to be an important part of the parent company until the late 1950s.

Clarence Iron Works, Port Clarence, 1920s. This image shows the blast furnaces of the ironworks and the offices directly in front of them. The coaling plant has been constructed which dates this picture after 1924. The square building to the left of the offices was the Engine House and, in the background to the right, more furnaces can be seen. (1588)

Although originally owned by Bell Brothers, the company merged with Dorman Long in 1899. This is a view of the Port Clarence site taken from Cochrane's Wharf in the 1930s. The vague outline of the Transporter Bridge can be seen, as can the mechanical charger on top of the blast furnace (right). (*5254 & 624*)

Clarence Foundry, 1920s. At one end of the foundry, men were 'ramming up' the moulding boxes, the process by which 'green sand' was packed around the pattern. Green sand was a mix of clay, sand and 'plumbago', an old name for graphite or black lead. Some men were 'floor workers' and can be seen creating their moulds in the ground. These workers were often paid a higher rate. After ramming up, crane weights (see bottom right with handles) would be put on top of the moulds before the molten iron was 'cast'. (*8763*)

Clarence Foundry, 1920s. At the other end of the foundry, workmen, including floor moulders can be seen working with different types of boxes. A sieve or a riddle can be seen on the left. The door was propped open with a cartwheel to let in additional light and air. In the foreground lie discarded cast metal shapes. Some boxes are steaming, indicating they have been recently cast. (*8762*)

These larger casings have been clamped up ready for casting ingot moulds (see p. 51). Because of their size, the casings were positioned below ground level so that the workmen could guide the overhead ladle to the casting hole. Another man follows behind, putting sand across the top to prevent flames. (*8764*)

Once the ingot moulds were cast and cooled, they were released from their casings. The hardened sand was then cleaned off by hand or using basic pneumatic tools. This image demonstrates the dirty and cramped working conditions of the men in the Ingot Foundry. (*8974*)

The pattern shop at Clarence Iron Works. All different types of skills were needed in an ironworks, including craftsmen who were used to working with wood. These men are making wooden patterns (see p. 19) which were placed in a mould and packed with sand. Once removed, mould was clamped together and hot metal poured into the shape of the item left in the sand. (*8760*)

Above: The machine shop at the Clarence Iron Works, where cast items were finished off. The boiler on the left was necessary to power the steam-driven sanders and lathes. (*8761*)

Left: In contrast to the manually charged blast furnaces of Ayresome Works (see p. 28), this furnace at Clarence Iron Works in the 1920s was mechanically charged. Iron ore, coke and lime were loaded into the top of the furnace automatically in specific quantities. There was no need for the lateral charging platform used by the older furnaces. (*8813*)

A view of one of the earlier versions of an open hearth steel furnace charging machine. The steel scrap, flux (lime) and other materials were placed in the container on the front of the arm and fed into the furnace through the door, which was raised by a pulley system. The charging machine moved on overhead tracks, similarly to a travelling crane, and the operator was able to manoeuvre it into position in front of any of the furnace doors. (*8812*)

Clarence Works Coke Oven – discharging side, 1920s. By 1925, over a million tons per annum of good quality coking coal was needed by the company's blast furnaces and 500 ovens had been built to ensure supply. Batteries of coke ovens were therefore a common feature on most sites. When the procedure was complete, a pusher ram forced the coke out of the discharge side, as is shown here. (*8992*)

Coke Ovens at the By-Products Plant of Dorman Long, Clarence Works. This image appears to have been taken during the First World War as some of the railway wagons have 'Ministry of Munitions' on the side. It is possible these wagons are bringing coal from the Durham coalfields. The towers at the back are for the separation of crude tar. Does anyone know what the water sprays were cooling? (*8993*)

A new modern Coaling Plant Wharf at Port Clarence was erected in the 1920s. At the time, it was able to provide ships with fuel for their own use (bunkering), or for freight, and could load and unload ships with up to 6,000-ton capacity. Vessels delivered raw materials to Middlesbrough and used the wharf to fill up with coal for the return journey, thereby increasing efficiency. (*2567*)

The wharves were equipped with electric and steam cranes which travelled on rails parallel to the dockside. This image, thought to be from the 1930s, shows the ability of the conveyor operator (housed in the cabin above) to extend the reach of the boom over the cargo holds of the ship. (*8795*)

By-products had always been a feature of Bell Brothers' portfolio since the days of Sir Isaac Lothian Bell. Uses were found for coal gas, waste heat (used to generate steam), crude tar, benzol, and various other by-products which were processed at the Recovery Plant (shown here) or at the Central Refinery. (*8804*)

North Ormesby

Alexander Brodie Cochrane came to Middlesbrough in 1854 and set up the Ormesby Iron Works, trading under the company name of Cochrane & Co. Primarily a pipe-making company in its early days, four blast furnaces were erected to provide a constant supply of iron to the foundry where the pipes were cast. From there, the pipes were taken to the stock yard near the extensive riverside wharf. The Pipe Foundry became a separate company in 1861 but continued to operate from the same site. The company also had shares in an ironstone mine at Stanghow and several collieries in County Durham. Cochrane & Co. became part of the Cargo Fleet Iron Co. at the end of the First World War but, in 1933, the furnaces, foundry and pipe making elements of the company were sold to the Stanton Ironworks Co. Ltd. It continued to operate as the Cochranes (Middlesbrough) Foundry Ltd until nationalisation but was closed by the British Steel Corporation in 1971.

Tees-Side Bridge & Engineering Works Ltd was a bridge-building and construction firm. Formed in 1898 by a syndicate led by Christopher Furness (see Cargo Fleet Iron Co.), the works was situated to the south of Cochrane & Co., between the railway lines and North Ormesby. With an expanding railway network, the Indian Government was an important customer in the company's early days and Tees-Side Bridge gained a reputation for bridge-building and steel frame construction. The company was forced to look for outside investment during the economic slump of the 1920s and was taken over by Dorman Long & Co. Ltd in 1930. During World War 2, the company was awarded several contracts for the War Ministry, including the construction of aircraft hangers for the Royal Air Force and landing craft for the Royal Navy. During the 1950s and 1960s, Tees-Side Bridge became a subcontractor of Dorman Long in the bridge and constructional side of their business. They were also involved in fabricating and installing key elements of the new blast furnaces installed at Clay Lane in the 1950s.

A modern view of Marsh Road, looking north towards the Navigation public house and Cargo Fleet Road. At one time, the land from the Cargo Fleet Road to Smeaton Street in North Ormesby was jam-packed full of buildings and construction yards belonging to the Tees-Side Bridge & Engineering Works (see p. 62). The company offices were on the right hand side immediately in front of the Navigation. *(J. H.)*

A view from Normanby Jetty looking west to Middlesbrough across the Graving Dock of the Tees Conservancy Commission about 1890. The entrance to Middlesbrough Dock can be seen, as can the blast furnaces of the Tees Iron Works (Gilkes, Wilson, Leatham & Co.). (*Original photographer unknown; printed from the original glass negative by Ian MacDonald.*)

Looking south from the wharf were the blast furnaces of Ormesby Iron Works (Cochrane & Co.), 1923–4. Cochrane & Co. made pipes of all dimensions and shipped them over the world, so having their own wharf was crucial. Railway tracks wound their way around the site, taking finished products to the stock yard (left). (*622*)

Looking west to the Transporter Bridge from Cochrane's Wharf, 1930s. The dockside cranes were used to transfer cargo from the ships, as well as to load pipes into the holds of the vessels. (*615*)

Casting tunnel segments in Cochrane's No. 1 Foundry, 1925. The ladle of iron is tipped under the control of the men operating the wheel on the right. The molten iron is poured into moulds which contain the shape of the pipe segments. (*613*)

Locomotive *Balkan* at Ormesby Iron Works, pushing a solo ladle, or possibly a slag pot. In the background a second locomotive can be seen, as well as the stacks of finished pipes in the stock yard. (*609*)

A wonderful image from about 1900 showing two men standing inside a large pipe or tunnel segment. Similar to the images from the Bridge & Constructional site, this technique was used to easily demonstrate the scale of the item. Perhaps the well-dressed man at the front is a member of the Cochrane family? (628)

Large Pipe Moulders, Ormesby Ironworks, 1890s. Back Row (left to right): Joe Batts; R. Stanton; J. Harris; -?- ; J. W. Williamson. Centre Row (left to right): S. Allport; L. Parker; J. Stanton; J. Longstafe; -?- ; A. Nash; John Onions (Under Foreman). Front Row (left to right): B. Harris; B. Kay; W. Harris; J. Price; J. Harris. Is anyone related to the Harris family who were well represented in this picture! (598)

Cochrane & Co., Ormesby Ironworks, 1910. Back Row (left to right): H. Fisher (1); J. Oliver; C. Onions; D. Whitehouse; H. Hill; W. Sheppard; A. Oliver; B. Watts; J. Onions (8). Centre Row (left to right): H. Allport; S. Allport; -?- ; W. Hardcastle; H. Nunn; C. Fisher; J. Onions; S. Elliott; P. Gorman. Front Row (left to right): -?- (with shovel); T. North; J. Brown; T. Reece. (595)

Cochrane & Co., Foundry Foremen, 1922–23. Back Row (left to right): W. Flavell; H. Bartholomew; J. Harper; E. Kemp; J. Harris; G. Robinson. Front Row (left to right): C. Fox; W. Goult; C. Greenwell; ? Edwin; J. Wilson; T. Whitfield. (592)

An elevated view of the stockyards and the rail tracks, looking south-west towards the blast furnaces. A picture such as this helps to establish the scale of Cochrane & Co.'s operation. (*639*)

Cochrane's Wharf, looking west towards Tees Iron Works with the river on the right. This elevated view shows the wharfside cranes and the neatly stacked pipes ready for loading. (*653*)

Works-affiliated sporting and social clubs were encouraged. This image of the Tees-Side Bridge & Engineering Works Cricket Club dates from 1923. Back Row: A. McSweeney. Second Row (standing left to right): H. Morley; F. Dyson; W. Lawson; F. Goat; W. Ennis; G. Fiddes; J. Pollitt; J. Gill. Front Row (sitting left to right): G. Grey; E. Husband; C. Hugill (Captain); H. Hurren (Secretary); A. Parkes. On ground (left to right): W. Pollitt; S. Spence. (*10361*)

Looking south along Marsh Road towards North Ormesby, 1948. The buildings on the left are the company offices. On the other side of the road were the railway sidings at the foundry. In the distance are the terraced houses of Smeaton Street and beyond those, Holy Trinity Parish Church in the Market Place. (*8224*)

Construction of the Mono-Rail, TSB, 1929. The sections of the rail were fabricated in the constructional shops and put together in the yard. The wall in the background is the Cargo Fleet Road. The Yorkshire Tube Works (Crewdson, Hardy & Co.) can just be seen through the structure. Below is a finished Mono-Rail. Below is a finished Mono-Rail, made for George Bennie, a Glasgow inventor, who launched his prototype 'rail-plane' in 1930. (*2059 & 2086*)

Left: In the 1920s and 1930s, Tees-Side Bridge was heavily involved in making gravity davits for lifeboats and galvanised pylons or towers like the one shown here. These structures were used as a form of advertising to anyone passing by on the Cargo Fleet Road. The strange thing was, Tees-Side Bridge had no waterfront access, so tests were carried out at Smith's Dock! The Yorkshire Tube Works is in the background. (*2071*)

Below: Large pipes loaded on flatbed bogies, 1938. The label on the side of the pipes reads 'Steel Work by the Tees-Side Bridge & Engineering Works Ltd, Middlesbrough'. (*2110*)

This image was taken in 1931 of workmen from Tees-Side Bridge supposedly working near Ormesby Crossing, which was at the southern tip of their site. It is a confusing puzzle of railway tracks but the photographer was looking north-west towards the Transporter Bridge. The Dock Clock can be seen on the horizon. (*2031*)

Part of the flood-defence scheme undertaken by Tees-Side Bridge in 1932, when the tracks leading down from the level crossing (Cargo Fleet Road) to the works were undermined. The culvert holds the waters of the Ormesby Beck, which flows behind the Navigation Inn. A temporary bridge was erected to carry the rail lines over the beck into the works. (*2068*)

An interior view of one of the machine shops at Tees-Side Bridge, taken in 1937. The machines are belt-driven lathes. On the post on the right was a box for goggles and instructions to wear them, but none of the men appear to have paid attention to this rule! (2125)

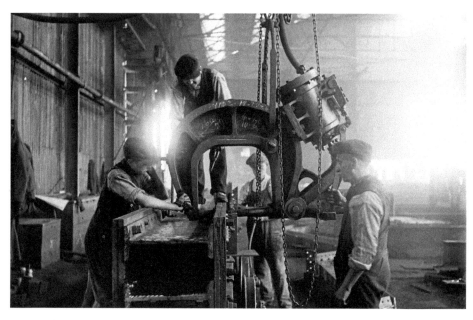

A cross-section view of a riveting machine at Tees-Side Bridge in 1929 being used to join sections of a girder together. This is a wonderful image showing the young lad trying hard to push with all his strength, the older man on top of the girder guiding it in to place and the experienced face of the older man, calmly operating the lever. (2011)

During the Second World War, Tees-Side Bridge took over the Cleveland Dockyards to construct Tank Landing Crafts (LCTs) for the Royal Navy. This image, from December 1942, shows LCT 715 (Mark 4 type) under construction in the Cleveland Dockyard. It was launched in May 1943 but lost off Normandy in June 1944. (*8008*)

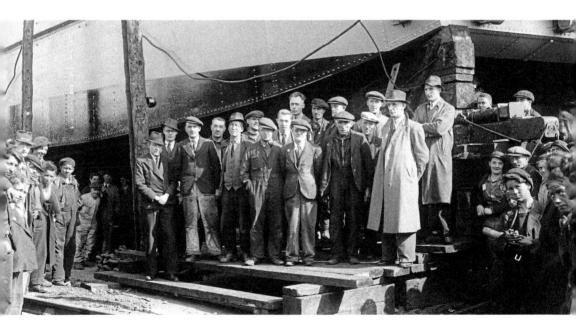

Taken at the launch of the 100th landing craft in 1943, this picture shows the foremen and workers. A similar photograph exists of the officials. More interesting perhaps are the on-lookers! Some were obviously workers in the yard, including the two women on the right-hand side, but it looks as though a few of the local schoolchildren have come along to watch. (*8093*)

In the 1920s, Tees-Side Bridge developed an 'iron bound composite roadway' plate which was laid a grid pattern in a concrete base. Tests were carried out at the corner of Cannon Street and Denmark Street in 1926 under the supervision of S. E. Burgess, the Borough Engineer and Surveyor. The group of on-lookers seems quite fascinated by the process. (*2147*)

A modern view of Marsh Road, looking south to Holy Trinity Church, North Ormesby. The terraced houses of Smeaton Street and beyond have gone, the Middlehaven Interchange dissects Marsh Road, and a solitary rail track is left as a reminder of what once was here. (*J.H.*)

Cargo Fleet

The Normanby Iron Works (Jones, Dunning & Co.) was opened at Cargo Fleet in the late 1850s, closely followed by Swan, Coates & Co., slightly further to the east in 1864. It was this latter company which would become the Cargo Fleet Iron Co. about twenty years later. In 1901, the whole of Cargo Fleet's share capital passed to the Weardale Steel & Coke Co. which, in turn, agreed a merger with the South Durham Steel & Iron Co. (SDS). Christopher Furness, the Chairman of SDS, saw that the site had potential. He brought in Benjamin Talbot to assist him, an engineer and inventor with a growing reputation who had recently returned from working in America. The site was radically overhauled; the old blast furnaces were demolished, the river frontage was extended and a new, 'integrated' works laid out, incorporating new blast furnaces, a steel plant, rolling mills, coke ovens and a coal washing plant. During the First World War, the company supplied the Ministry of Munitions with shell steel for casings, a move that prompted the company to build a new Head Office on the Middlesbrough Road (see p. 76). After a shaky start, the fortunes of the company improved under Talbot's management; he remained at the helm until his death in 1947. In the early 1930s, the company considered a merger with Dorman Long & Co. Ltd, but it was called off. In the early 1950s, financial control of Cargo Fleet passed its sister company, SDS, although it continued to trade under its own name. Despite heavy investment in the company throughout the 1960s, nationalisation was to signal the end for Cargo Fleet. Nearly two thousand workers lost their jobs when the British Steel Corporation moved steel production to Lackenby in 1973. Although parts of the works and offices remained for more than a decade, the gates were finally shut in the late 1980s.

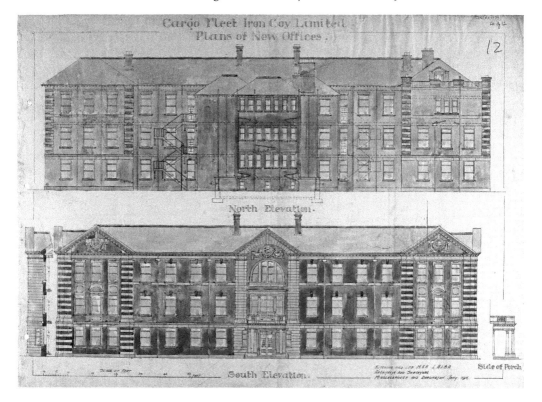

One of the original architectural drawings for the new Cargo Fleet Offices by Kitching & Lee, dating from 1916, which are held at Teesside Archives. (*TA DC.ES 17.12.4*)

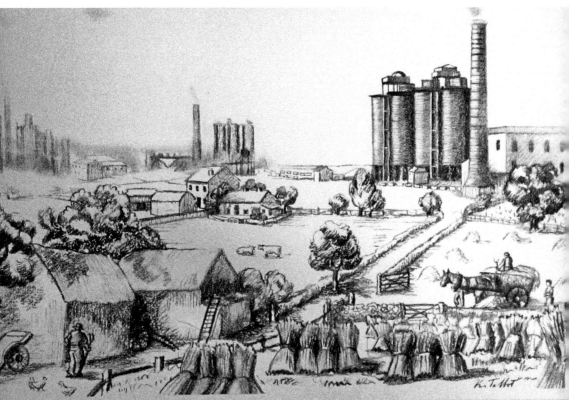

This old print must have been used as the inspiration for the Viva Talbot lithograph (see below), although she has used artistic license to add more romantic illustrations of farming life than are shown in the original. It shows the blast furnaces of the Cargo Fleet Iron Co. (on the right), the Normanby Iron Works (centre) and the Ormesby Iron Works (far left) as they were in around 1865. The land immediately to the south of Cargo Fleet was undeveloped until the early twentieth century, when Sir Christopher Furness initiated a radical overhaul of the site. Viva Talbot was the daughter of Benjamin Talbot, who was instrumental in helping Sir Christopher Furness turn the fortunes of the company around. *(1599; M. Meredith-Hardy; J. H.)*

The description on the original glass negative was: 'Ship "Muskoka" in graving dock. Stern View' and dates from November 1891. The *Muskoka* was build by Richardson, Duck & Co. of Stockton-on-Tees. The blast furnaces of Cargo Fleet Iron Co. can be seen in the background, at that time owned by Swan, Coates & Co. (*Photographer Unknown; Printed from the original glass negative by Ian MacDonald.*)

'Graving Dock repairs. View of the south wall', November 1891. Another view of the Cargo Fleet blast furnaces in the distance, courtesy of this wonderful photograph. The charging platform on the furnaces is visible, as is a locomotive pulling wagons across an elevated rail track. (*Photographer Unknown; Printed from the original glass negative by Ian MacDonald.*)

This postcard has no date but the two furnaces mean that it must date from after the Furness/Talbot overhaul of the early 1900s (see also p. 73). The large pipe captured the gas created in the blast furnace process and transferred it to the washers to be cleaned of dust prior to re-use, the first time that had been achieved. (*J. H.*)

Originally, it was intended to build five furnaces at Cargo Fleet but only two had been built at the time of this image. The mechanical chargers were imported from America, where Talbot had worked until 1900. Behind the furnaces was the new Humboldt coal-washing facility and the coke ovens. (*2615*)

The gantry above the storage bunkers was designed so that the trucks of ironstone, limestone and coke could pass along it by gravity alone. 7,000 tons of ironstone per week from the company's Liverton mine was needed to feed the furnaces. (*2611*)

Left: Part of the battery of 100 Kopper coke ovens at the Cargo Fleet Iron Co., around 1905. When the coke was discharged from the ovens, it fell by gravitation into a special truck which transported it to a conveyor which, in turn, fed the blast furnace. It was advantageous to have coke ovens on site, especially if there was also a steel works, as the gas could be used to fuel the open hearth furnaces, thereby saving on coal. (*2614*)

Below: The Humboldt Coal Washer and Conveyor, around 1905. Coal was automatically drawn up from a bunker by conveyor to the coal washer and then ground at the rate of 60 tons per hour before being transported to the coke ovens. (*2617*)

A view of the blast furnaces in the background, the Humbolt Coal Washer conveyor gantry and the coke oven battery at Cargo Fleet, 1905. (*2623*)

Workers from the Cargo Fleet Iron Co., 1952. Do you recognise anyone? (*1593*)

The aerial photograph shows the Cargo Fleet Iron Works looking north-west. The works covered a large expanse of land from the Middlesbrough Road to the river, but note how undeveloped the fields in the foreground were (now an industrial estate). The row of terraced cottages on the right was called Forest Place. The new Head Office, started in 1916, was designed by local architects, Robert R. Kitching and William Lee. The design for the original building (see p. 69) shows three gables, whereas the current building has four. When the works were closed down in the late 1980s, it lay empty for several years before Redcar & Cleveland Council used it as offices. However, when the Council moved out, the building fell into disrepair and was a magnet for looters and vandals. Thankfully it was saved from demolition by a local firm, Python Properties, which has redeveloped the property and has included a heritage gallery on the ground floor. (*10359; J.H.*)

South Bank, Grangetown and Eston

It was Bernhard Samuelson & Co., later of the Newport Iron Works and Britannia, who built the South Bank Iron Works when he first came to the area in the early 1850s. Although three furnaces were constructed initially, this number had doubled by 1866. The furnaces were located parallel to the tracks on the north side of the N.E.R. railway from Middlesbrough to Redcar. Samuelson sold the South Bank works to Bolckow, Vaughan & Co., and Thomas Vaughan, son of John Vaughan, was heavily involved in their development. By 1869, a further three furnaces had been erected.

Messrs T. L. Elwon & Co. erected three blast furnaces in 1853 and 1854 which they named the Cleveland Iron Works but they sold this site almost immediately to Bolckow & Vaughan. Thomas Vaughan was again involved in their development and by 1869, when the Staffordshire & Worcester Institute of Mining Engineers made a visit to the site, there were five furnaces. From the top of the 95-foot-high furnace the visitors were afforded a very fine view of the countryside and the river. Messrs. Elwon, Malcolm & Co., then built a new iron works near Eston Junction and called it Clay Lane, due to its proximity to a farm of the same name. However, by 1869, this works has also become part of Bolckow & Vaughan's empire and was operating six blast furnaces.

Between 1853 and 1876, the original Eston Ironworks disappeared under Bolckow & Vaughan's new Cleveland steel plant constructed to the east of the Branch line. Thousands of skilled workers and labourers flocked to the rapidly growing towns of South Bank and Grangetown. When the British Association visited the Eston Steel Works in 1889, they described it as follows: 'Covering a vast extent of ground, and lit up at night by the silvery electric light they present a unique appearance, and constitute one of the "sights" of the neighbourhood. The works now comprise three hematite blast furnaces, four 8-ton Bessemer converters, six 15-ton basic converters, seven open-hearth steel melting furnaces...' and various different types of mill. The works were producing an average of 5,000 tons of steel per week and employing 13,000 men. Such a population required extensive facilities and, in addition to the civic buildings erected, there was provision made for sporting activities, such as the Ironopolis Cricket & Football Ground near Branch End and allotment gardens.

During the 1920s, Bolckow & Vaughan acquired the Darlington Rolling Mills, the Eston Sheet & Galvanising Company and agreed to merge with Redpath Brown, an Edinburgh-based steel construction company. This latter acquisition was to make Bolckow & Vaughan more competitive in the structural steel construction side of their company, but the economic depression at the end of the 1920s prompted the merger with Dorman Long & Co. Ltd. It was Dorman Long which planned and implemented the much needed modernisation of the Cleveland, Clay Lane and South Bank sites in the 1950s. They demolished all the old furnaces which were, by that time, between fifty and eighty years old, and installed larger, more efficient, modern furnaces and a new coke oven plant. The wharf at South Bank was extended and modernised so that larger vessels could dock and be unloaded more efficiently. In 1967, the manufacturing side of the business was nationalised and the British Steel Corporation took over the site. The ensuing shrinkage from Middlesbrough benefitted the area for many years, until the demise of the industry in the 1980s. Only the wharf and South Bank Coke ovens are still operating; all steel making and manufacturing moved to Lackenby and Redcar.

It seems fitting that this book should end with images of where it all began – Eston. The closure of the Eston mine in 1949 signalled the end of an era – an era that triggered the growth of this region and led to Middlesbrough and Teesside being known the world over for their association with the iron and steel industry.

This undated photograph has been identified as the South Bank Iron Works, probably from about 1900. There are eight blast furnaces and at least two hoists to the charging platform. The rail wagon identifies it as part of Bolckow, Vaughan & Co. (*1607*)

South Bank Iron Works, the blast furnaces in 1954. The configuration of the furnaces appears to be different but it is difficult to tell from the image how many there were at this time. In the foreground are slag pots on solo bogies. (*9076*)

In 1956, work began on the demolition of the South Bank blast furnaces. The old Cleveland furnaces can be seen in the background. This was No. 1 furnace being demolished on 18 April. The demolition of Nos 2, 3 and 4 furnaces followed and they were all gone by the end of the year. (*9141*)

Larger and more efficient furnaces were built at Clay Lane between 1953 and 1961. This image shows No. 1 furnace (left) still under construction, but Nos 2 and 3 had been in operation for several years. (*10101*)

Above: South Bank Iron Works – the last 'tap' of Ferro-Manganese from No. 2 Furnace in 1955, prior to its demolition. The man on the left with the shovel is ensuring the sand bank is high enough to hold the molten metal. Compare the safety clothing (or lack of it) to the man in the second image. (*9106*)

Left: Cleveland Works, Clay Lane, 1960. A worker modeling new Italian-made safety clothing specifically for blast furnacemen. I wonder if it ever caught on? It must have been incredibly hot and difficult to move properly in such bulky clothing. (*10083*)

The new Otto Simon Carves Coke Ovens at South Bank under construction in August 1954. It was anticipated that these new ovens would carbonise 23,000 tons of coal every week. These ovens were demolished in 1973 and replaced with the Willpute Coke Ovens which are still in use today. (5757)

Just over two years later, in November 1956, the first oven was 'pushed' at the new facility. Everyone turned out to watch; there were workmen peering over railings from the roof and around the coke guide to see the action, as the hot coke erupted from the oven and fell into the coke car below. Some of these coke cars were manufactured at Tees-Side Bridge & Engineering. (5749)

The Dorman Long tower and the gas holder at South Bank Coke Ovens in 1957. In the distance the dockside cranes of South Bank Wharf can just be seen. The Dorman Long tower was built as a coal supply bunker for the Coke Ovens. The gas holder's function was to act as a reservoir of coke oven gas to fulfill demand as it was needed by the steel mills and other power needs. It also acted as a pressure controller to increase stability across the distribution system. (*10328*)

The Dorman Long tower still stands as an instantly recognisable landmark. Although its original purpose as a storage bunker is no longer required, there are two huge water tanks at the top containing 1,000 tons of industrial water. These are connected into the fire-fighting hydrant ring main supply. The current gas holder replaced the original which was struck by lightning in 1971. (*J.H.*)

Right: South Bank Wharf, probably in the early 1940s. Unloading ore at the wharf of Cleveland Works using the old method of discharging the ship. Ore was loaded into hoppers and transferred to rail wagons, guided in by a workman. This must have been an incredibly laborious process. (*5555*)

Below: By the end of the decade, South Bank Wharf had been updated and huge new ore unloaders were installed. In August 1949, the SS *Aldebaron* delivered 10,000 tons of African manganese ore to South Bank. At 475 feet long, she was the longest ship until that date to dock at South Bank. (*10308*)

These signs illustrate and emphasise the combined histories that Dorman Long carried with it until nationalisation in 1967. The Cleveland & Clay Lane sites had been owned by Bolckow, Vaughan & Co. until the two companies amalgamated in 1929. (*5256 & 5644*)

The blast furnaces of Cleveland Iron Works about 1900, when owned by Bolckow, Vaughan & Co. There were six furnaces in varying stages of blast and three hoists to raise the raw materials to the charging platform. The engine house is the square-shaped brick building to the right of the picture. The Cleveland Works furnaces were demolished in the 1950s, when the entire site was restructured. (*1604*)

These basic Bessemer Convertors were built as a result of the Gilchrist Thomas experiments in the late nineteenth century and were operational for about twenty-five years. The converter was an open-topped egg-shaped vessel, supported on 'trunnions' which allowed it to pivot to receive raw materials and again to discharge the molten metal. The metal can be seen pouring down into a waiting ladle. (*2257*)

A locomotive steaming away from 'A' Furnace and tapping the blast furnace (bottom) at Cleveland Iron Works. These images probably date from between the 1930s and 1950s. Cleveland Iron Works was located just to the north-west of Grangetown, with the Steel Works directly north. The 1890 issue of Bulmer's Directory notes that '...almost the entire population is employed in and about the extensive works of Messrs. Bolckow, Vaughan & Co., Limited'. (*2267 & 5286*)

Cleveland Steel Works, 1947. A team of men working together at an open hearth furnace, either to clear the tap hole from blockages, or to take a sample. The sample had to be taken from the hottest part of the furnace and the metal bar used was very long and heavy. (9004)

Slag Disposal Grounds, Cleveland Works, 1957. The slag pots were pushed by a diesel locomotive and discharged liquid slag. The process was not automated, so the driver was required to pull the chock to turn the pots. (9208)

By 1959, Dorman was operating two steel plants at the Cleveland Works site. This photograph was taken in the North Steel Plant and shows the process of 'teeming' steel into hot top ingots. The operator controlled the stopper in this bottom-pouring ladle which was raised and lowered to control the metal. (9293)

An important role with any steelworks was that of the Roll Turner. These men, pictured in 1954, are calibrating and refurbishing rolls for the mills to ensure they performed to their full potential. (9069)

Above: A view of the Cooling Banks at the Medium Section Mill. As the sections exited the finishing mill they passed to the cooling banks. The banks move forward very slowly and the rails are cooled by natural convection. The workmen in this image seem to be painting something on each section. (*9233*)

Right: This is a view of a small-diameter rod emerging from a mill. As it did, the handler caught the end of it with tongs or pincers and looped it around his body in preparation of feeding it back through the mill. The young lad was probably an apprentice. (*9128*)

The demolition of the old Cleveland Iron Works furnaces in 1957. The newly erected furnaces can be seen behind. It is only when they are seen together that it can be appreciated just how large the newer furnaces were. These new furnaces were, in turn, demolished in the late 1980s. (*9239*)

The newer furnaces were installed from 1954 onwards. This image shows the shuttering for the foundation of one of the blast furnaces. (*7881*)

Right: As part of the site clearance at Cleveland and Clay Lane, the Blast Engine House and Power Station were demolished. These had been built in 1864 and 1900 respectively by Bolckow, Vaughan & Co. This image shows a dumper-truck driver looking up at the dating stone of the old power station. (*9059*)

Below: Clay Lane – a general view of the site showing the blast furnaces and the new work started in 1953 to reorganize the site. The South Bank furnaces can be seen on the left, with the Ferro Manganese furnaces to the far right. (*9043*)

A familiar sight to many in 1960 when this picture was taken – the main office buildings at Cleveland Works, overshadowed by the gas holder. The Pay Office was at the rear of this building. (*9342*)

Dorman Long opened an apprentice training centre at Cleveland Works in the early 1960s. This image, dating from 1962, shows the apprentices using different tools and machines under the supervision of instructors (in white coats). (*9454*)

Eston Mine, May 1949, looking north. The building on the left was the hauling engine house. Ironstone wagons were ready to enter the mine. In the middle distance were the workmen's cottages of California. On the horizon, the blast furnaces of the Cleveland Works can be seen, as can the shape of the Dorman Long tower and the gas holder. (*9783*)

16 September 1949, the final day of work at Eston Mine. Mr Russell, Under Manager, and a colleague uncouple the last set of wagons. (*9785*)

16 September 1949, the final day of work at Eston Mine for this group of miners. They received their last pay packet that day. (*9788*)

16 September 1949. The last shift of men leaving the Drift Mouth as Eston Mine was closed. It had been continually worked for ninety-nine years. (*9789*)

Primary Sources

Teesside Archives, Middlesbrough, British Steel Collection [Cat. Ref: BS].

Bodleian Library, University of Oxford, Oxford, Trade Catalogue of Dorman Long & Company Limited, 1924. John Johnson Collection: Ironmongery 9 (9).

South Durham Steel & Iron Company and Cargo Fleet Iron Company, *Steelmaking Illustrated* [An album of sixteen lithographs by Viva Talbot] (Printed in Middlesbrough, undated).

Secondary Sources

Aberconway (Lord), *The Basic Industries of Great Britain* (London, 1927).

James, S. (2013) *Growth and Transition in the Cleveland Iron and Steel Industry, 1850 to 1914.* Doctoral Thesis, Durham University. Accessed at: http://etheses.dur.ac.uk/6957/

Lillie, W., *The History of Middlesbrough: An Illustration of the Evolution of English Industry* (Middlesbrough, 1968).

Marley, J. 'Cleveland Ironstone', in *North of England Institute of Mining & Mechanical Engineers Transactions*, vol V., 1856-57, page 165.

Mountford, C. E. and D. Holroyde, *The Industrial Railways & Locomotives of County Durham, Part 1* (Melton Mowbray, 2006).

Pollard, A. J. (Ed.), *Middlesbrough: Town and Community 1830–1950* (Stroud, 1996).

Robertson, W., *Middlesbrough's Efforts in the Great War* (Middlesbrough, 1920).

Tighe, T. R., *Tees Side Bridge: The Rise, Fortunes and Dissolution of a Private Company* (Middlesbrough, 1980).

Willis, W. G., *South Durham Steel and Iron Co. Ltd* (Middlesbrough, 1969).

Yasumoto, M., *The Rise of a Victorian Ironopolis: Middlesbrough and Regional Industrialization* (Woodbridge, 2011).

Maps

Cleveland & Teesside Local History Society, *Middlesbrough's History in Maps* (Middlesbrough, 1980).

Ordnance Survey (The Godfrey Edition), *Yorkshire Sheets: 6.10 (North Middlesbrough, 1913); 6.12 (South Bank & Grangetown (North), 1927); 6.14 (Central Middlesbrough, 1913); 17.01(Eston, 1927).*

Ordnance Survey 6 inch, Yorkshire (1853)

Ordnance Survey 25 inch, Yorkshire VI. (1895 and 1929).

Electronic Sources

Nineteenth-century British Library Newspapers (Accessed via Gales News Vault).

Times Digital Archive (Accessed via Gales News Vault).

Bulmer's *History and Directory of North Yorkshire* (1890). Accessed at: http://www.genuki.org.uk/big/eng/YKS/NRY/Eston/Eston90.html

Permission to reproduce images and maps is gratefully acknowledged from:

- Teesside Archives: Images from the British Steel Collection have their identification number in brackets after the caption. Also: Map of the Ironmasters' District Ref: BS.DL.7.5.2.1 and Architectural Plan of Cargo Fleet Offices Ref: DC.ES.17.12.4.
- Bodleian Library, University of Oxford: Frontispiece map from Trade Catalogue of Dorman Long & Company Limited, John Johnson Collection: Ironmongery 9 (9).
- English Heritage: Aero Pictorial Image Ref: B81857 (AFL03) – reproduced on p.7 under ID 2571.
- M. Meredith-Hardy: Lithograph entitled 'General View of Cargo Fleet Works in 1864' by Viva Talbot.
- Ian MacDonald (photographer): Images shown on p.57 (top), 71 and 72 (top) were printed by Ian from the original glass plate negatives; original photographer unknown.
- My photographs bear my initals.

Acknowledgements

This book would not have been possible without the assistance of all the digitisation volunteers who have worked with me at Teesside Archives since 2009. My thanks to (in alphabetical order): Paul Boden, Andy Colley, Jean Donnelly, Paul Fagan, Mick Jones, Geoff Lofthouse, Colin Robinson, Peter Tait and Chris Twigg. Since 2011 Paul B., Jean, Colin and Peter have continued to volunteer – thank you.

The staff at Teesside Archives have been very supportive in permitting access to the British Steel Collection and in applying their considerable collective knowledge to assist me. In particular, I would like to thank Helen Kendall (Conservator), Ruth Hobbins (Principal Archivist), Janet Baker and David Tyrell (former Archivists) and the current team: Stuart, Cori, Michelle and Kimberley. I am also grateful for the generosity of academic colleagues and interested members of the public alike who have willingly shared their knowledge about Middlesbrough's iron and steel industry with me, in particular, Dr Stephen James of Teesside University. Morale support came from my friend and mentor Professor Tracy Shildrick, whom I cannot thank enough.

I owe a great deal of thanks to all the team who worked with me on the British Steel Archive Project: Dr Jenny Search (Community Engagement Manager), Simon Sheppard (Archivist), Tony King and Zoe Stewart (Conservators), Angela Whitecross, Sally Sculthorpe and Rachael Kenny (Access & Education Officers), Karen Oxley (Project Administrator) and Tosh Warwick (PhD student).

Despite the difficulties in the global economy over the past few years, steel is still being produced on Teesside by Sahaviriya Steel Industries (SSI) UK Ltd. The 'moth-balling' of Redcar's blast furnace in early 2011 cast a dark cloud of despair and gloom over the whole community, only lifted when it was restarted by SSI just over a year later. The closure served to reinforce the importance of promoting, protecting and celebrating the iron and steel heritage of this area. Over the years, I have been grateful for the support of local politicians and representatives of Corus (now Tata Steel) and the Community trades union in getting that message across. Senior managers at SSI (UK) Ltd have also been incredibly generous with their time. In particular I would like to thank John Baker (formerly Corus TCP, now SSI (UK) Ltd) Thanks also to Colin Agar and Dr Mike Copeland of SSI (UK) Ltd, and Edward Bilcliffe and Stuart Duffy from William Lane Ltd, Middlesbrough.

Most importantly, I want to thank my partner, Jan Norman, whose ancestors came to South Bank from Ireland to work in the iron industry, for her love and support. Right here...with you...

Author's Info

Dr Joan Heggie is a Research Fellow at Teesside University in Middlesbrough. From 2008-2011 she managed the British Steel Archive Project and has spent many years encouraging people to use the British Steel Collection, as well as promoting the importance of Teesside's industrial heritage to wider audiences. She is currently researching the art of Viva Talbot and is writing a book about the Talbot family home, Solberge Hall. Joan lives in Saltburn by the Sea with her partner, Jan.